**Vitali Ponomarenko**
**Nikolay Pushanko**

**Ejection devices in mass transfer processes of sugar industry**

Vitali Ponomarenko
Nikolay Pushanko

# Ejection devices in mass transfer processes of sugar industry

LAP LAMBERT Academic Publishing

**Impressum / Imprint**

Bibliografische Information der Deutschen Nationalbibliothek: Die Deutsche Nationalbibliothek verzeichnet diese Publikation in der Deutschen Nationalbibliografie; detaillierte bibliografische Daten sind im Internet über http://dnb.d-nb.de abrufbar.

Alle in diesem Buch genannten Marken und Produktnamen unterliegen warenzeichen-, marken- oder patentrechtlichem Schutz bzw. sind Warenzeichen oder eingetragene Warenzeichen der jeweiligen Inhaber. Die Wiedergabe von Marken, Produktnamen, Gebrauchsnamen, Handelsnamen, Warenbezeichnungen u.s.w. in diesem Werk berechtigt auch ohne besondere Kennzeichnung nicht zu der Annahme, dass solche Namen im Sinne der Warenzeichen- und Markenschutzgesetzgebung als frei zu betrachten wären und daher von jedermann benutzt werden dürften.

Bibliographic information published by the Deutsche Nationalbibliothek: The Deutsche Nationalbibliothek lists this publication in the Deutsche Nationalbibliografie; detailed bibliographic data are available in the Internet at http://dnb.d-nb.de.

Any brand names and product names mentioned in this book are subject to trademark, brand or patent protection and are trademarks or registered trademarks of their respective holders. The use of brand names, product names, common names, trade names, product descriptions etc. even without a particular marking in this works is in no way to be construed to mean that such names may be regarded as unrestricted in respect of trademark and brand protection legislation and could thus be used by anyone.

Coverbild / Cover image: www.ingimage.com

Verlag / Publisher:
LAP LAMBERT Academic Publishing
ist ein Imprint der / is a trademark of
OmniScriptum GmbH & Co. KG
Heinrich-Böcking-Str. 6-8, 66121 Saarbrücken, Deutschland / Germany
Email: info@lap-publishing.com

Herstellung: siehe letzte Seite /
Printed at: see last page
**ISBN: 978-3-659-47763-8**

# CONTENTS

# Introduction

Ejection devices had a wide spread occurrence in various branches of the industry, including food industry. Their simplicity and reliability work attracted the attention of practitioners, but there are serious shortcomings that do not allow us to use their capabilities to the fullest. Such deficiencies primarily there is a lack of theoretical justification of mass transfer processes that occur in them, not fully resolved the question of hydrodynamics, low efficiency.

As research, ejection devices appear more design that is sophisticated is explained hydrodynamics flow, mass transfer, expands the boundaries of the equipment.

This work is devoted to research in the sugar industry on using of ejection devices for carrying out different processes: activation of lime with milk, liming, carbonation sulphitation, de ammonization, carbonation gas purification from impurities.

Copyright certificates, patents, some of the works being widely adopted in sugar factories, some in the implementation phase, protect each work of the authors.

## 1.  Area of using of the ejection devices in the food-processing industry

Implementation of the various physic - chemical processes, food processing materials due to mass-transfer processes , movement and mixing of solid, gaseous or liquid phase. The most widely used well-established methods of mechanical and hydrodynamic processes intensification based on the local energy input. One of the simplest is to treat products by ejection method. The advantages of the ejector are no moving parts, ease of construction and maintenance. In addition to the simplicity of design of these devices, a high intensity of mixing, mass transfer, heat transfer. However, using of the ejection processing equipment products requires careful learning of this problem, the development of methods for calculating various types of ejectors. Ease of implementation in technological schemes of various food industries, the reliability of the augur well for their widespread use.

An ejector is a device wherein there is a transfer of kinetic energy from one medium (working) moving at high-speed area in the potential energy of the mixed flow (intake of working and - ejection). Job of ejector obeys conservation of energy in the form of the Bernoulli equation: created in the convergent section of the flow low pressure working environment, which causes suction flow in a passive medium, which is then mixed with the active working environment, thus there is a transfer of kinetic energy from the working environment to passive and subsequent alignment of energies.

In general, the ejector consists of working nozzles, inlet chamber, the mixing chamber and diffuser. Efficiency such ejector, however, does not exceed 30%. Improving of ejection devices led to the creation of liquid-gas ejectors with elongated mixing chamber. In such devices the length of the mixing chamber 30 reaches a diameter of 35 ... of the mixing chamber [1], and a suction volume ratio of the apparatus is increased by about two times.

Ejection device except the pump function perfectly fulfills the function compact and efficient mass transfer equipment. In the mixing chamber through a constantly updated contact, surface interfacial phase is an intensive exchange of substances between the liquid and gas phases. In this regard, the ejection devices as equipment, combines the advantages of ejectors and the possibility of mass-transfer processes, are increasingly used in various sectors of the food industry, in particular - sugar.

In the sugar industry, several processes can be effectively carried out in intense ejection devices.

Among them:

- mixing phase components (sugar and lime juice) for the subsequent process of defecation sugar solution;
- implementing processes sulphitation barometric water, juice and syrup;
- execution of the process the first and second carbonation;
- De-ammonization of condensates;
- Cleaning of carbonation gas from harmful contaminants and dust.

Consider the features of these processes. Mixing process of diffusion juice with milk of lime in a sugar factory neglected. It is believed that the introduction of lime milk supply duct directly into the diffusion juice in the liming is sufficient for carrying out processes in the chemical purification unit, since there is no destruction of mucous precipitate, which formed on the preliming step.

However, obtaining juice with high purity from the non-sugars may be provided uniformly distributed throughout the lime treated solution volume for chemical reactions possible deposition of non-sugars, wherein the activity of the milk must be high.

In order to intensify the mixing raw juice and lime with the activation of the last sugar factory was proposed to use supercavitators, take advantage of cavitation effects for grinding undissolved particles of lime, lime activation and intensive mixing. However, such devices operate long since destroyed themselves and cavitation effects.

One of the main ways to improve the process of sulphitation water and sugar solutions is to develop a theoretically based stimulation techniques that reduce fuel sulfur. Becoming increasingly important to making full use of $SO_2$ from sulfitation gas in terms of sugar production from the point of view of the intensification of mass transfer processes, and reduce harmful emissions.

An example of replacing metal-intensive, low-productivity equipment for irrigation type sulphitation products of sugar production [2] are proposed in the 70s of last century ejection sulphitation devices. Such a replacement has significantly intensify the process of mass transfer of sulfur dioxide to normalize the sulphitation off all equipment, improve the manageability of the process. Furthermore, the high degree of utilization of sulfur dioxide in the proposed hardware ejection greatly reduces emission into the atmosphere of gas [2]. Prolonged operation of the ejection equipment sulphitation products of sugar production also identified some shortcomings of his work. These include low efficiency of spray gun, labor services gas communications due to their frequent clogging, and the fragility of exhaust communications.

For the process of the first and second carbonation (chemisorption process between $CO_2$ and CaO to form calcium carbonate high adsorption capacity), recently suggested the use of high ejection apparatus, as the first stage carbonator.[3] At the same time intensifying the chemisorption processes in the first stage of saturation to give a high degree of supersaturation of $CaCO_3$, which is a condition of mass nucleation of crystalline calcium carbonate high adsorption capacity and their subsequent growth in the bubble stage saturator.

Using of ejection devices to comprehensively address the first and second carbonation unknown.

De - ammonization of condensates in a sugar factory neglected. Using of de – ammonization ammoniums as condensates of extraction feedwater had positive effect on the extraction plant and evaporation plant juice. This water has no hardness, decontaminated from microorganisms. The concentration of ammonium in ammonium condensates for their possible using in the diffusion does not exceed 50 ... 80 mg/l.

To achieve a concentration of ammonium in the condensates mainly bubbling method of de - ammonization or de - ammonization in film mode to the packed column. The desorbing is used as the air and water vapor. In the latter case, for carrying out the desorption process takes about 3 % steam to beet. In the case of using air as the desorbing per cubic meter condensates must be 150 ... 200 $m^3$. The process is very energy consuming de - ammonization to the present time there is a search for optimal circuits and equipment for its implementation.

For carbonation gas cleaning of contaminants and dust, and its subsequent cooling are applied in the form of large equipment or the cyclone chamber and the collecting scrubbers for cleaning and cooling. Analysis of the results of studies of this process has shown that the precipitation of various impurities in the carbonation gas has its optimum temperature at various purge gas. Efficiency gas cleaning from dust in large equipment there is always a lower quality of the process due to the presence of stagnant zones, imperfections flow hydrodynamics.

These processes of the sugar industry may intensify considerably when used as highly efficient mass transfer apparatuses ejection equipment. Such devices are known favorably to the simplicity of design, can work in a wide range of parameters of the gas, let you easily adjust the workflow have low metal, provide a high intensity mass transfer processes. Their production in the workshops of sugar factories for their own means no difficulties.

A common drawback of constraining the use of ejection apparatus in the sugar industry is their low coefficient of ejection, the lack of reliable results on the kinetics of mass transfer processes, which does not produce reliable estimates of the equipment.

## 2. Mass transfer in devices with dispersed jet fluids

Intensification of mass transfer is the subject of considerable scrutiny. With the accumulation of knowledge are opportunities to create new high-performance equipment for processes that were carried out in the traditional bubble or spray devices.

All known methods of intensification of mass transfer processes aim at the highest possible increase in the interface or the effect on this surface by various physicochemical methods.

Let's briefly examine the theoretical foundations of various methods of mass transfer of matter from one phase to another and the main factors that affect the absorption of liquid gas.

### 2.1 Models of the mechanism of conduction

The first attempt to explain the mechanism of mass transfer was made by Nerst. According to his theory on a solid surface, which is dissolved in a moving fluid, there is a stationary liquid film thickness $\delta$, in which the mass transfer occurs only by molecular diffusion [4, 5].

Lewis and Whitman continued development of this theory, and expressed the view that on the two mobile phases, between which the transfer of components, there are two films that are adjacent to each of the phases and are the diffusion resistance of each phase. If we consider the diffusion in one direction z, perpendicular to the direction of fluid flow y, the convective diffusion equation has the form:

$$\frac{\partial c}{\partial t} = D \frac{\partial^2 c}{\partial z^2} - w \frac{\partial c}{\partial y} \; , \tag{1}$$

With boundary conditions:

$$\beta\Delta= -D({\partial c}/{\partial z})_{z=0}, \tag{2}$$

where, $t$ - time, $w$ - velocity of the medium, $\Delta = C_p - C$ - the driving force, $Cp$ $and$ $C$ - concentration of the components at the interface and in the bulk phase, $\beta$ - mass transfer coefficient.

Within the substance of the film is carried by the diffusion flux, the phase interface and offers no resistance to her equilibrium concentration components. The process is stationary. Convective term is neglected and the mass transfer equation takes the form:

$$D\frac{\partial^2 c}{\partial z^2} = 0 \tag{3}$$

Integration with the boundary conditions and film thickness $z_0$ allows you to find $\beta$:

$$\beta = {D}/{z_0} \tag{4}$$

Disadvantages of these theories is that they do not take into account the hydrodynamic interaction between the phases, mass transfer coefficients indicate the dependence of the diffusion coefficient of the first power, which is not always confirmed by practice.

Further development of the theory of mass transfer was obtained in Higbee [4,5]. Considering the industrial gas-liquid contact apparatus, which are operated under short-term contact of the two phases, it is speculated that this contact time enough to reach steady state. When a gas bubble in the fluid rises, the liquid film, which is in direct contact with the bubble, is updated in a time equal to the time length for lifting its bubble diameter.

In solving the equation (1) is considered acceptable and the convective component distribution phase contact time constant for all surface elements.

As a result, the equation obtained mass-transfer coefficient kg in liquid phase in the form of:

$$k_r = \sqrt{{4D}/{(\pi\tau)}}, \tag{5}$$

where $D$ - diffusion coefficient, $\tau$ - the duration of contact between the phases.

Mass transfer coefficient, according to the theory Higbee depends on molecular diffusion coefficient in 0.5 degree, which is confirmed by experimental data of foreign researchers [4] that the exponent of the diffusion coefficient varies in the range of 0.5 - 0.75.

Higbee complement theory began from works of Dankverst and Stabnikov (surface renewal theory), who took the time phase contact unlike Higbee not the same, but different. Their assumption was made that the new elements are in contact with the medium at different time intervals - from zero to infinity. The distribution function of the surface elements on the residence time in contact is:

$$\Phi = se^{-st}, \qquad (6)$$

where $\Phi$ - the probability that an element of the surface will be a period of time $t$, before being replaced by the new element of the liquid from the flow core, $s$ - refresh rate of the surface.

Mass transfer coefficient in this case is:

$$k_\Gamma = \sqrt{Ds}, \qquad (7)$$

Kishinevskiy M. took the time to contact phases similar Higbee - constant and mass-transfer coefficient found expression similar to (5), but unlike the others suggested that the contact time for the mass transfer occurs both turbulent and molecular diffusion. This ratio is called the effective diffusion coefficient:

$$D_{e\phi} = D + D_t, \qquad (8)$$

where $Dt$ - turbulent diffusion coefficient.

A brief overview of the major theories of mass transfer shows that they are designed for systems with a steady movement phases. Just a theory update involves not stationary mass transfer time update phase contact surface.

Considered the main mechanism of mass transfer explains the transfer of matter from one phase to another after the formation of the interface that is formed under the surface (eg, a liquid droplet, which has already been formed by the decay of the jet at the nozzle exit nozzle). Assessing the impact of mass transfer with the direct formation of a new surface treatment of the theory does not take into account because of the difficulty quantifying. However it is known that the formation of new surface speed transfer of matter from one phase to another is higher than in steady contact phases.

## 2.2 Mass transfer during the formation of the contact surface

In reality, significant influences on mass transfer have transient non-stationary processes preceding the formation of a stationary substance. Such processes are spraying liquid: droplet formation at the outlet of the atomizer; knocking against hard partition; coalescence and crushing drops when they collide.

These processes cause intense update phase contact surface [6]. In these cases, mass transfer processes by 1 - 2 orders of magnitude higher than at steady motion [6, 7, 8, 9].

When calculating the appropriate equipment these processes relate to the transition and are not always taken into account. This approach greatly simplifies the engineering calculation designed equipment and provides a somewhat exaggerated its size.

Accounting for mass transfer processes during the formation of droplets or bubbles is complex and unsolved problem to date, offered various explanations for the mechanism of mass transfer intensification. In [9] provides an overview of some of these works.

Authors [10] explains the acceleration of mass transfer due to turbulence in the interface and the formation of her during the droplet growth regions with different surface tension. The presence of the difference of the surface tension on the surface causes the formation of vortices. Submitted Schlieren photographs confirm the qualitative picture, but for quantifying they cannot be used.

Jet fluid flow regime because of its small residence time refers to the case when you need to take into account the final effects. Calculation of the mass transfer is carried out as a penetration model on the assumption that the penetration depth is less than the radius of the substance droplets. In this case [6] nA material flow onto the surface of the drop can be found from the expression:

$$n_A = \left(c^* - c_0\right)s(t)\sqrt{D/(\pi t)}, \qquad (9)$$

where $c^*$, $c_0$ - equilibrium and the initial concentration of dissolved substances in the droplet , $D$ - diffusion coefficient ; $s$ - surface area of the droplet , which increases in size as a result of joining the masses in time $t$.

Having a second fluid flow through the cross section of the capillary:

$$Q = \pi d_0^2 u_0 / 4,$$

where $d_0$ - capillary diameter ;

$u_0$ - velocity of the fluid out of the capillary,
in time $t$ is equal to the amount of drops:

$$v(t) = \frac{4\pi r^3(t)}{3} = Qt.$$

We define the area of the drop surface at any moment of its formation t, expressing the last expression of the droplet radius:

$$S(t) = 4\pi r^2(t) = 4\pi \left( \frac{3d_0^2 u_0}{16} \right)^{2/3} t^{2/3}. \tag{10}$$

The surface area of droplets formed $s_f$ during its formation $t_f$ will:

$$S_f = 4\pi R_f^2 = 4\pi \left( \frac{3d_0^2 u_0}{16} \right)^{2/3} t_f^{2/3}, \tag{11}$$

where $R_f$ - radius formed drops.

Multiplying and dividing the expression (10) $t_f^{2/3}$ and given by the expression (11):

$$S(t) = 4\pi R_f^2 \left( \frac{t}{t_f} \right)^{2/3}. \tag{12}$$

Material flow into the drop during the formation of drops:

$$N_A = \int_0^{t_f} n_A dt = \frac{24\pi}{7} \sqrt{\frac{D t_f}{\pi}} R_f^2 (c^* - c_0). \tag{13}$$

Full flow of matter $N_A$ can also be defined by the expression:

$$N_A = (c - c_0) \frac{4}{3} \pi R_f^2, \tag{14}$$

where $c$ - concentration in the gas stream.

Equating the expressions (13) and (14) for the flow of matter can obtain expressions for the degree of extraction :

$$\varphi_t = \frac{c - c_0}{c^* - c_0} = 1.45 \sqrt{t_f}. \tag{15}$$

The latter expression shows, that the degree of extraction of the substance during the formation of drops by 1,45 times higher than at steady state, the mass transfer.

Calculation of mass transfer under this formula does not take into account the effect of substance transfer by convection during the spreading of the liquid in the droplet during its formation.

Accounting convective effect radially Ilkovych was performed [11]. The coefficient of the degree of extraction agents during the formation of drops differs from the Higbee factor 1,52.

Thus, it has been proved theoretically [6] that at the time of the droplet mass flux exceeds the transfer of the same substance in steady flow. When you create a device which will be repeated surface renewal, we can achieve a significant increase in mass transfer and ignoring boundary effects lead to significant errors in the calculations.

The described mechanism of mass transfer in the formation of droplets is also highly simplified and does not take into account a number of effects. For example, it is known that the formation of droplets at the atomization fluid nozzle precedes the occurrence of surface waves on the liquid film when it is out of the nozzle after the nozzle, but the ripple effect is not considered. In devices that operate under a dispersed phase droplet collisions occur repeatedly, leading to repeated contact surface renewal phases and increases the flow of material, but it is also not considered in the above theories.

Besides these phenomena, which occur during atomization liquid droplets in homogeneous environment and the complexity of quantifying mass transfer when there are also difficulties in mass transfer phenomena accounting system polydisperse droplets.

## 2.3 Mass transfer among polydisperse droplets

Liquid stream, which flows from the spray nozzles, under the influence of external forces breaks up into droplets of different sizes, ie the system of polydisperse droplets formed with Gaussian size distribution. In such a system drops mass transfer implemented through various mechanisms. In very small droplets [4, 12], the transport occurs by molecular diffusion. In the case of very large drops [4, 13] instead of the laminar circulation observed hydrodynamic regime, similar to the intense internal mixing. Mass transfer occurs due to molecular and convective diffusion. In the intermediate-sized droplets occurs laminar toroidal internal circulation, which reduces the length of the path in the process of molecular diffusion. For example, the droplet size of 1 mm is significantly slower absorbing gas per unit surface area than larger droplets. With a diameter of 4,2 mm drops [12] inside it by convective diffusion carries 35 - 40 % hydrogen fluoride.

After absorption of poorly soluble gases when a significant proportion of the mass transfer resistance is concentrated in the liquid phase, it is desirable to use such dispersants fluid which allow to obtain is quite small droplets (with a mass transfer surface is large), or vice versa, is very large (4 - 5 mm) [5]. In this case it will be best achieved for renewal phase contact surface, and hence the conditions improve gas absorption.

In [14, 15, 16, 17] that one of the main conditions for effective operation of the equipment in which the absorption is carried out in the spray mode, the organization is fine and monodisperse liquid atomization, as well as providing uniform distribution of the liquid density of the dispersed phase in the cross section unit. In papers of Vyskrebtsov V.B. given experimental confirmation of these provisions [18,19].

Based on analogy between heat transfer and mass transfer processes, an example of heating with a different length cube edges and removing them warm thermal diagrams found that the temperature cube (edge length 0,06 m) is a monotone increasing (the same as too large droplets, and the whole liquid mass). Cube with edge length 0,02 m heats much faster, and its temperature exceeds a constant equilibrium temperature of the entire system. Then there is a cooling of the dice by transferring heat to a less heated.

On the basis of the experiments of Viskrebtsov V.B. concluded that small droplets absorb absorbent much faster than the large and the entire mass of the liquid as a whole. The smaller droplet, greater the maximum concentration in which this gas reaches in the process of absorption of the drop, maximum value reaches before the concentration. Therefore, in gas absorption polydisperse droplets will be considerably slower (due to the desorption of gas from the fine droplets, and then removals larger) than in the case of a monodisperse spray.

Thus, a polydisperse droplets in the system, which is formed by dispersing the liquid, there is an ambiguity estimation mechanism of mass transfer. Consider all these nuances in general impossible. Therefore, in many cases, the use of volumetric mass transfer coefficient and conclude on the effectiveness of the equipment on the average characteristics.

## 2.4 Influence of a steam stream on mass transfer

Despite the complexity of the mechanism of mass transfer in a dispersed mode in the system of polydisperse droplets there is a difficulty quantifying the influence of the vapor flow in the gas phase.

This part flow of the reason that in almost all cases, the gas from which the absorption of the target component is not in thermodynamic equilibrium with the liquid, i.e. is not saturated with water vapor at the process temperature. In this case there are two mutually opposite flow:

- The flow of target component in the gas phase adsorbate;
- The vapor stream from the surface of the drop in gas environment.

Taken separately components of the process of mass transfer in detail in several papers. Since the absorption, the authors considered process in [4, 5, 6, 9, 18, 19, 20]. Vaporization processes in detail in [4, 6, 9, 21]. Considerably less work considering the mutual influence of one mass flow to another.

Effect of vapor flow on mass transfer of $CO_2$ by the example of carbonation process sugar solutions in the spray mode.

One promising area of physic - chemical treatment of non-sugars from sugar solution is to conduct the process in a two-section saturator [22, 23] with the first stage of phase contact spray pressurized bubble and subsequent phases with a total cocurrent. This unit receives a high quality juices, achieved more complete utilization of $CO_2$ carbonation gas (the main components of the gas is a mixture of carbon dioxide at a concentration of 30 ... 35 %, water vapor and inert gas - nitrogen $N_2$).

By proceeding in a carbonation unit remain outstanding some aspects of mass transfer of carbon dioxide in the sugar solution in the opposite direction of diffusion of water vapor in the carbonated gas.

In the first stage saturator - spray absorber - centrifugally dispersed jet nozzle sugar solution with a concentration of lime 2.5 - 3% CaO reacts with carbon dioxide carbonation gas which enters the absorber is saturated with water vapor until an equilibrium partial pressure of the cooling water at a temperature of about 30° C and a temperature of 75 ... 80° C, after compression in the compressor. In the spray absorber of carbon dioxide diffuses to the interface followed by a rapid chemical reaction of the second order and simultaneous saturation of carbonation gas with water vapor to equilibrium temperature of the sugar solution $77^0$ C (Fig. 1) . You could say that: the spray absorber component mixture of $CO_2$ and $H_2O$ under pressure $P$ at temperature t in the film thickness $\delta$. Gases flow towards each other through two limiting surface, and inert $N_2$ gas is supplied and is not removed from the film.

Communication flow of carbon dioxide ($N_{CO2}$) and water vapor ($N_{H2O}$) in inert gas ($N_2$) with concentrations at the boundary of the film and its properties of the system of equations [4]:

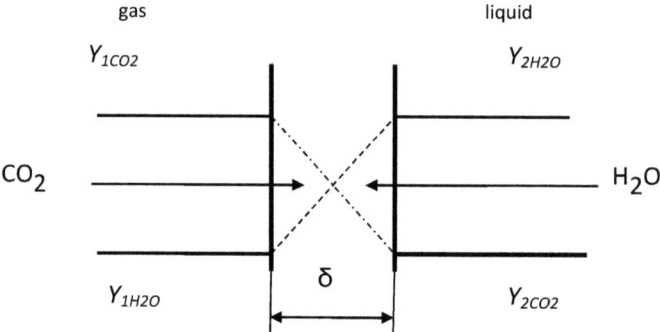

Fig. 1 Streams components diffusing through the film

$$\begin{cases} \dfrac{N_{CO2}}{D_{CO2.N2}} + \dfrac{N_{H2O}}{D_{H2O.N2}} = \dfrac{P}{RT\delta} \ln \dfrac{Y_{2N2}}{Y_{1N2}} \\[2em] N_{CO2} + N_{H2O} = \\[1em] = \dfrac{D_{CO2,H2O}P}{RT\delta} \ln \dfrac{\lfloor (N_{CO2}+N_{H2O})/N_{CO2} \rfloor Y_{2CO2} - \lfloor (N_{CO2}+N_{H2O})/N_{H2O} \rfloor Q Y_{2H2O} + Q - 1}{\lfloor (N_{CO2}+N_{H2O})/N_{CO2} \rfloor Y_{1CO2} - \lfloor (N_{CO2}+N_{H2O})/N_{H2O} \rfloor Q Y_{1H2O} + Q - 1} \end{cases} \quad (16)$$

where:

$$Q = (1/D_{CO2,H2O} - 1/D_{CO2,N2})/(1/D_{CO2,H2O} - 1/D_{H2O,N2}) \quad (17)$$

Diffusion coefficient of component $A$ to component $B$ is calculated as follows [24]:

$$D_{A,B} = \dfrac{10^{-3} T^{1.75s} [(\mu_A + \mu_{B)}/\mu_A \mu_B]^{0.5}}{P[(\sum \upsilon_A)^{1/3} + (\sum \upsilon_B)^{1/3}]^2} , \quad (18)$$

where    $\mu$ - molecular weight components

$\sum \upsilon$ - molecular diffusion volumes.

When performing the following calculations, it was assumed that the carbonated gas during his stay in the absorber time to get enough steam to a state of equilibrium. To confirm this hypothesis were conducted assessment calculations by three independent methods. Obtained similar results time gas saturation with water vapor in the range of $9,5 * 10^{-3} - 5,2 * 10^{-3}$ s. It should also be noted that the calculations were made with the assumption of net evaporation influence oppositely directed flow of $CO_2$ was not considered.

The system of equations obtained by the method of successive approximations [4]. Find flow ratio of components for a certain characteristic time at which the value of the mole fractions are constant on both boundary surfaces and are:

$$Y_{1CO2} = 0,273; \quad Y_{1H2O} = 0,025; \quad Y_{1N2} = 0.702 \text{ - on first surface}$$

$$Y_{2CO2} = 0,238; \quad Y_{2H2O} = 0,288, \quad Y_{2N2} = 0.474 \text{ - on the second, and the}$$

following conditions:

- The temperature of the liquid in the absorber 77°C;
- The temperature of the inlet gas to the absorber - 73°C,
- Partial pressure of water vapor in the gas $p^o{}_{H_2O}- = 0,035 * 10^5$ Pa,

which corresponds to the temperature of the inlet gas to the compressor and an absorber 27°C;

- Gas outlet temperature 76°C, the saturation pressure $p^H{}_{H_2O} = 0,408*10^5$ Pa;

- The gas pressure in the absorber $P = 1,39 * 10^5$ Pa .

These conditions correspond to the state of the carbonated gas inlet to the spray absorber and its output.

Up 7% solution of (16) gives the ratio of water vapor flow in a stream of carbon dioxide in the form:

$$N_{H2O} = 5,21* \ N_{CO2} \tag{19}$$

The latter relation leads to several important conclusions:

- carbonation gas saturation of water vapor in the spray absorber slows oppositely directed flow of $CO_2$;

- also holds the following statement: the rate of absorption of carbon dioxide with a sugar solution is significantly reduced due to the evaporation of water.

Virtually all processes heat - and mass flow of the reactants except the title , which are directly involved in the process, there is concomitant flows of components that do not participate in physical and chemical processes of the target, but will affect the main stream. This takes place, as in this example, when the partial pressure of water vapor in the gas phase partial pressure below the equilibrium vapor pressure at the temperature of liquid. In such cases, it is necessary to consider the effect of the vapor flow as target of absorber transfer process slows down. To eliminate the negative impact of the reagents necessary to prepare in advance to participate in the mass-transfer processes - to bring them into balance in terms of that slow the target process (at saturation sugar solutions in the spray mode - carbonated gas is saturated with water vapor and an intensive process of evaporation, which slows absorption dioxide sputtered carbon sugar solution). If this is not done, the unit volume of the reactants come to equilibrium, and the longer it takes, the greater the deviation from this condition. Target mass transfer process while slowing.

Based on the foregoing , a method carbonation sugar solutions [25], which comprises the carbonation gas saturation water vapor partial pressure until the gas equal to $0,85 * 10^5$ ... $1,4*10^5$ Pa, which corresponds to the temperature of carbonation gas 95 ... 110 ° C.

Carbonation gas saturation water vapor is especially true during the second saturation at $90 - 102^0$ C.

At saturation carbonation gas with water vapor to these limits are accelerated absorption of carbon dioxide under the influence of unidirectional vapor flow (Stefan flow). Furthermore, the passage of gas through such a carbonation sugar solution is cooled it becomes satiated and water vapor are formed in the water droplets with dissolved carbon dioxide, which also contributes to the increased use of carbonation gas $CO_2$.

## 3. Ejection devices in sugar industry

### 3.1 Equipment for sulphitation products of sugar production

In the production of sugar beet and sugar cane is mandatory process sulfitation - processing sugar solutions with sulfur dioxide. Distinguish processes sulphitation water, juice, syrup, sugar klerovki (brown sugar solution) - raw. The main requirements to the equipment in which the process sulphitation is the execution of the purpose of this process to the maximum of $SO_2$ to reduce the harmful effects of the gas on the environment.

The goal of treatment sulphitation different products is different, however, for its implementation can be used the same type of machine. More detailed information on this subject can be found in the literature [26, 27].

Historically, the processes sulphitation sugar solutions implemented in sulfitators irrigation type liquid- gas.

Sulphitation gas for processing sugar solutions produced by burning sulfur in the ovens, followed by sublimation at its vapor cooling of sulphitation gas deposition of suspensions in settling chambers.

Equipment for the production and purification of Sulphitation gas had a small operation reliability (especially sublimator) dimensions cooling chambers reached enormous proportions, and, nevertheless, the need for process sulphitation forced sugar producers to work on such equipment.

Finding ways to address these problems led to sulphitation products of sugar production in the liquid-liquid (sulfitation using acid, which was added in the required quantities to sulphitation solution). Equipment for the process sulphitation in this case is a simple mixing apparatus for mixing two liquids.

The disadvantages of this method include increased aggressiveness acids and risk of its storage, the presence of special acid tanks.

Sulphitators for products of sugar production should ensure low power consumption with high efficiency of mass transfer processes, be simple in design.

An example of replacing metal-intensive, low-productivity equipment for irrigation type sulphitation products of sugar production in liquid- gas has proposed 70-ies of the last century ejection of sulphitation installation [2].

Their advantages over irrigation devices such as:
- A significant intensification of mass transfer processes;
- Ten times less specific metal devices;
- Significant reduction in sulfur dioxide emissions into the atmosphere;
- Saving of sulfur on the implementation process sulphitation.

The equipment of sulphitation typical installation includes:
- Incinerator lump sulfur and produce sulfur dioxide;
- Sublimator sulfur vapor deposition;
- Jet apparatus with a cyclone to separate the gas-liquid phases.

Of the equipment listed above is the least workable sublimator sulfur. It requires periodic cleaning of the internal surfaces of pipes deposited glassy and fairly solid sublimed sulfur. As practice shows, such equipment fails very quickly, disrupting operation mode of sulphitation installation.

Model jet ejection device is a short mixing chamber and the nozzle as a working fluid is used to disperse the disc with three holes are formally inkjet nozzles. This device has a number of disadvantages:
- the inability to maintain an optimal pH of the solution at the ever-changing flow of fluid;
- low utilization of sulfur dioxide, as evidenced by a corroded exhaust pipe.

It should be noted that the presence of sulfur sublimed on the cooled walls of the pipes the machine indicates its incomplete combustion in the stove. One reason may be a lack of air during combustion.

Analysis of gray combustion furnaces in the production of sulfuric acid shows that more complete combustion of sulfur in the combustion zone, an additional air supply.

Combining afterburning sulfur on the wall with additional air supply and purification of suspensions we developed, implemented in many factories cyclone converter sulfur, which successfully replaces sublimator. Moreover, if the allocation of the sulfur in the tubes sublimator it indicates incomplete combustion, leading to overrun, the cyclone afterburner sulfur deposition is not observed. It is all converted into the gaseous state and is reacted with the processed fluid.

Refinement of sulphitator design is to replace the system with a liquid dispersion of the jet mode to disperse, lengthening the mixing chamber to the optimum, as well as additional turbulence flow liquid-gas emulsion directly in the mixing chamber .

Consider a typical work station sulphitation (but without sublimator) in the classical situation when working sugar factory.

In the case of stable operation of the plant, and hence liquid flow ejectors create vacuum of equal magnitude in its gaseous communication hence work almost equal.

Terms of the group of sulphitators significantly change when changing fluid flow in any of them. From the theory of operation of ejection apparatuses known that the amount of the ejected gas flow rate essentially depends on the fluid flowing through the nozzle. In this case, there may be cases when the flow decreases sulphitate liquid ejection data by gas of sulphitators will not be enough.

Fig. 2 shows an example of a typical situation in a sugar factory in a constantly changing flow rate. Sulphitator number one running at rated speed, when it serves a full flow of liquid and gas pipeline supplying a vacuum of 30 Pa, taking into account losses in the pipes in the manifold vacuum of 20 Pa. When this gas is ejected enough to maintain the desired pH of the liquid.

Sulphitator number 2 operates at a low flow rate of the liquid with underload and develops cravings only 10 Pa. In this case, taking into account losses in such piping sulphitator cannot provide the required ejection amount of sulphitation gas to maintain the desired pH of the liquid at the outlet of sulphitator. With these parameters, work station sulphitation inevitable inversion of the gas flow in air communication of second sulphitation.

Above example refers to sulphitation station, the main element of which is the ejection apparatus with compact liquid jets. Coefficient ejection apparatuses such can reach 10 [1].

Thus, in a constantly changing flow rate to maintain a stable pH values in each sulphitator, which regulated by the technological mode of a sugar factory, it becomes very problematic.

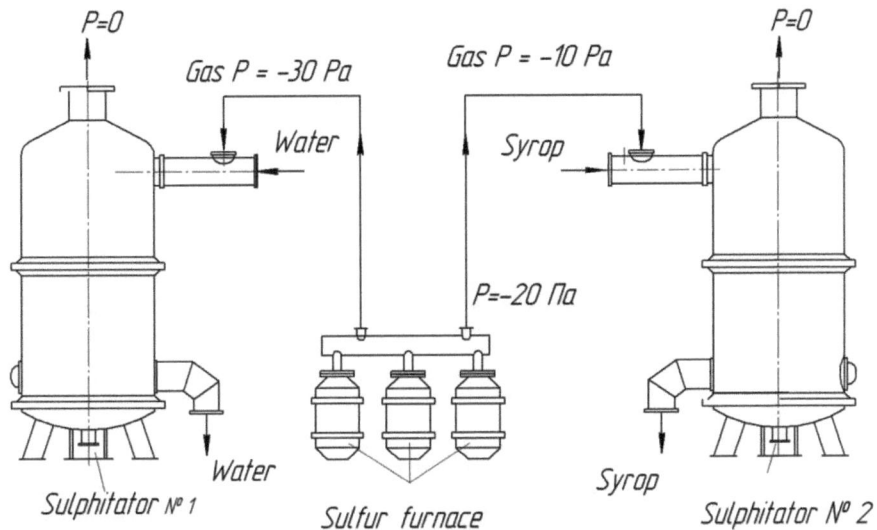

Fig. 2 Typical operating circuit of sulphitators

To solve this problem when you create such conditions of the ejector in which will be greatly enhanced ejection coefficient, and will be increased before vacuum sulphitator. This is possible if the ejector will not work when the jet mode of liquid outflow into the mixing chamber, while dispersed expiration mode. [1]

Fig. 3 shows the same station in the same sulphitation production situation. Only ejection coefficient and thus a vacuum in the supply line of sulfur dioxide of sulphitator increased 10 times, and to maintain the normal combustion of sulfur in gas furnaces Communications in order to re-burning the unburned sulfur and purification from impurities installed cyclones with great water resistance. Therefore, the gas manifold vacuum persists necessary.

Fluctuations in flow of liquid before sulphitator and in case of increase in gas communication dilution ten times at the entrance to the first created by sulphitator Rod 300 Pa, and the second sulphitator even when it is unstable in the Rod 100 Pa. However, even for such negative pressure sufficient ejection amount of sulfur dioxide required for the regulated pH of the liquid exiting from each sulphitator.

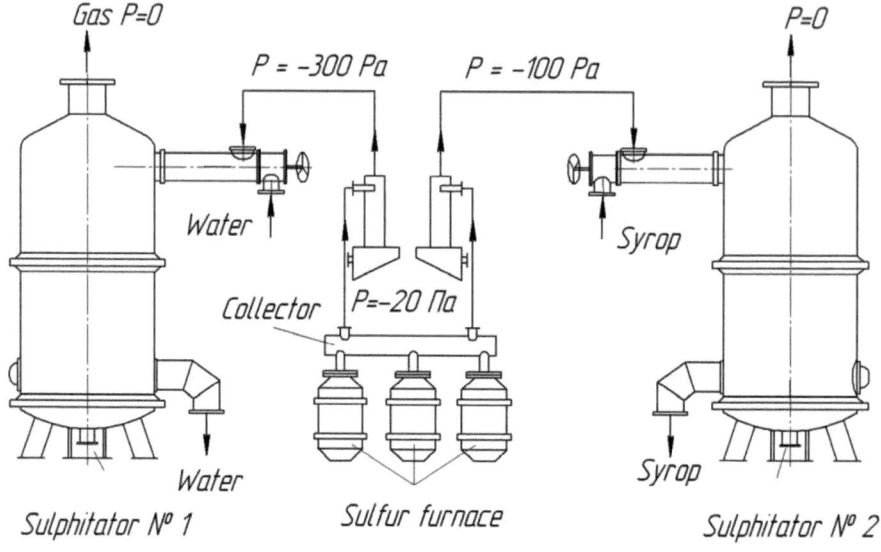

Gas P=0

P = -300 Pa

P = -100 Pa

Water

Collector

P=-20 Па

Syrop

P=0

Sulphitator № 1

Sulfur furnace

Water

Syrop

Sulphitator № 2

Fig 3. Scheme 3 sulphitation station after reconstruction

Expected result from the introduction of improved sulphitation installation practices and certified to ensure the reliability of its work on changes in fluid flow range of 50-150 % of nominal, increase utilization $SO_2$, in reducing fuel sulfur achieving stable pH value of the liquid.

Figure 4 shows the improved scheme of sulphitation barometric water, which works on some sugar factories in Ukraine.

It includes gray combustion cyclone furnace with afterburner sulfur, which insulated and equipped with additional air supply. Sulphitator of barometric water is an ejection device with an elongated mixing chamber. On one side of a pressure of 0.2...0.4 MPa centrifugal jet nozzle disperses water and barometric served by sulphitation purified gas and the other side of the mixing chamber tangentially adjacent to a cyclone separator emulsion. Sulphitated water removed from the bottom of the separator and the flue gas through the central exhaust pipe removed to the

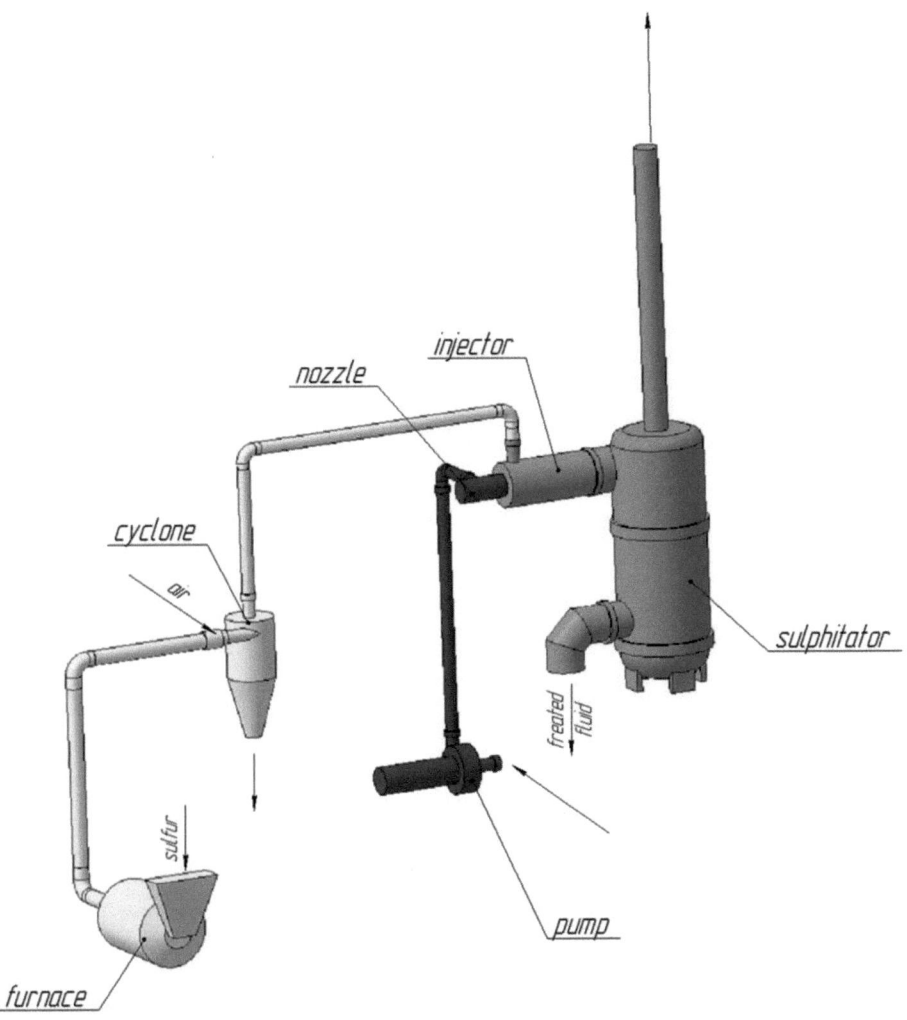

Figure 4 Station of sulphitation barometric water

atmosphere. Installation work stably when the liquid flow rate in the range of 50-150 % of nominal, can reduce fuel sulfur technical 20-25 %.

Dwell on design features of sulphitator and nozzles for spraying liquid in the mixing chamber.

### 3.2 New constructions of nozzles for centrifugal-jetting dispergation of fluids

For the main element of the ejection, system-working nozzle - has developed a number of special designs centrifugal spray nozzles with large flow passages, providing the opportunity to work in contaminated liquids. Such nozzles differ filled spray, high flow coefficient. Calculation of the main nozzle size is not given here, if necessary it can be done using the method [28].

Here is a description of several, in our opinion, the most successful designs. A nozzle according to certificate of authorship SU 1382499 (Fig. 5) with inclined feeders, axial outlet and plug used for dispersing liquids with the possibility of regulation of its flow by axial movement of the plunger [29].

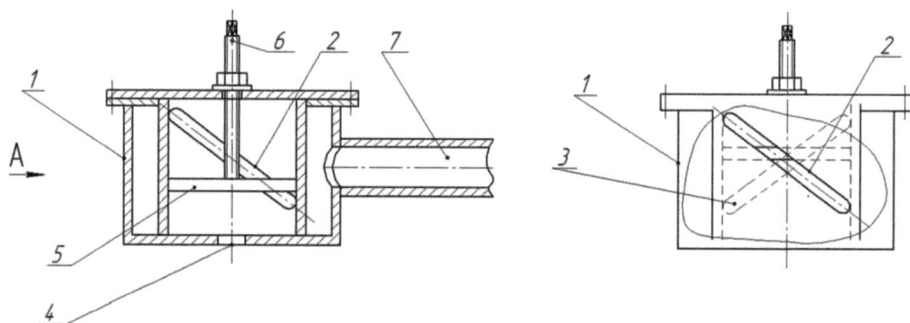

Fig. 5 adjustable nozzle

Injector 1 comprises a cylindrical chamber with side slotted feeders 2 and 3 and the axial outlet 4. The chamber 1 is mounted axially movable plunger 5 to the lead screw 6. Pipe 7 is intended for supplying a liquid.

The nozzle operates as follows.

Fluid under pressure introduced into the intermediate space of the chamber 1 through the pipe 7 and then introduced into the chamber through slotted twisting nozzle passages 2 and 3.

The nature of the fluid in the chamber 1 is dependent on the twisting position of the plunger 5 with respect to the slit channels 2 and 3. In the lower position of the plunger 5 open area of the channels 2 and 3 is small and close to the arm twisting chamber radius. At the same flow, rate minimum and maximum degree of spin liquid. At the outlet formed hollow spray pattern characteristic for a centrifugal fluid outflow. When lifting the plunger open area of the channels 2 and 3 increases, the degree of twist of the liquid decreases and increases flow rate , and filling it with the spray increases. Mode liquid outflow is centrifugal jet.

In the upper position of the plunger 5 Channels 2 and 3 are fully open. The liquid, which enters the chamber 1 through the bottom tightening section of the canal 2 and 3, acquires a rotational movement in one direction and the liquid, which enters the chamber 1 through the tightening piece upper channels 2 and 3, acquires a rotational motion in the opposite direction. As a result, the total angular momentum is close to zero, the expiry of the liquid spray nozzles becomes the characteristic of jet nozzles, and flow rate reaches a maximum.

Use of this nozzle extends the operating range of the heat - mass transfer equipment, ejection apparatus by increasing the range of liquid flow control.

At the bottom position of the plunger nozzle works as centrifugal and has shortcomings of the nozzles of this type: hollow spray pattern, low coefficient of flow rate.

Using the nozzle according to the invention patent UA 99671 (Fig. 6) for the atomization of liquid increases the efficiency of spraying the liquid at the bottom of the plunger position: atomization of the liquid becomes a centrifugal jet [30].

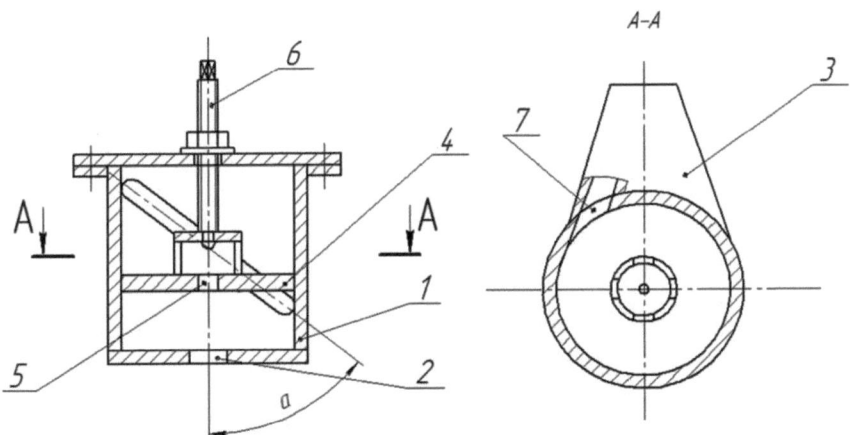

Fig. 6 nozzle with a central axial hole

Injector 1 comprises a cylindrical chamber with an axial outlet in an end face 2 of the chamber 1. It abuts the side surface of the chamber 3 of the liquid feeding branch pipe, which end shaped slit whose edges are connected tangentially to the inner surface of the chamber 1 on either side of the nozzle axis. The chamber 1 is mounted axially movable plunger 4, wherein an axial hole 5. The plunger 4 can be moved along the axis of the nozzle by rotating the screw 6.

Side slit duct 7, the fluid supply to the nozzle body 1 is situated in a plane inclined to the axis of the chamber 1 at an angle of 30-60 °, and directed tangentially to the inner surface of the chamber 1. Specified interval angle of the feeding channels 7 is selected from the consideration that if the angle exceeds 60 °, then to fully open the channels requires a significant stroke of the plunger 4. This leads to difficulties in regulating the flow of liquid within a wide range, and increase the dimensions of the nozzle. When the tilt angle is less than 30 $^0$ channel 7 small stroke of the plunger 4, thus failing to achieve the required accuracy regulation.

In the lower position of the plunger, formed therein an axial hole, allows to increase the uniformity of liquid distribution over the cross section of the torch by supplying fluid through the central axial hole in the movable plunger. This improved distribution of the liquid at the bottom position of the plunger is possible, because the presence of the axial bore there through with an upper cavity of the liquid atomizer is flagged as an axial jet to the outlet nozzle. As the fluid enters the chamber and twist by a nozzle hole with a significant angular momentum, the nozzle hole in the main interaction occurs tangential flow and axial flow of liquid that passes through the axial hole in the plunger. This leads to an interaction of the two streams , the equalization of speeds , some of the momentum transfer from the swirling flow axial and vice versa, and as a result, a uniform distribution over the cross section of the liquid flame spraying, which is typical for centrifugal jet nozzle.

When the upper and middle position of the plunger axial fluid flow out of the hole in the plunger does not affect the nature of liquid outflow from the injector nozzle.

In the case of relatively stable flows of fluid ejection, apparatus is recommended to use the nozzle (Fig. 7) on the patent for utility model UA 45125 [31]

The nozzle consists of a cylindrical chamber 1 twist , which is tangentially connected to the outlet 2 , is placed within the chamber dividing wall 3 twisting with axial opening 4 , with one end of the twisting chamber holds injector nozzle 5, and at the other end of the chamber closed by screwing cap 6.

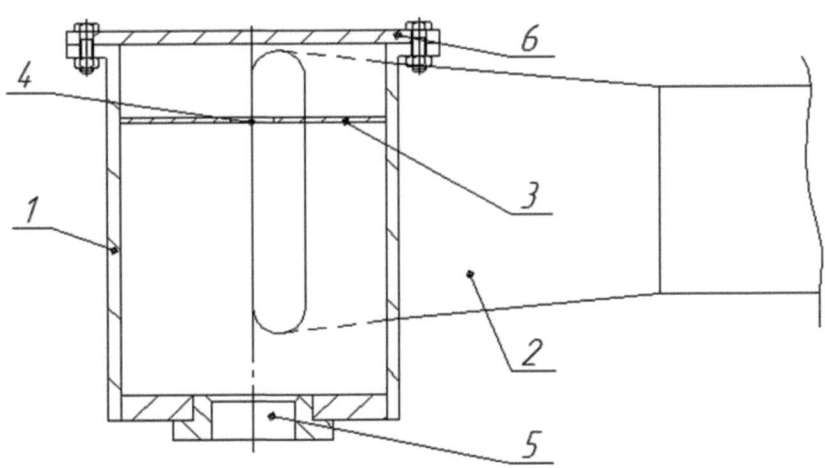

Fig. 7 Unregulated centrifugal spray atomizer

Placing baffles in the twisting chamber should be such that it overlaps no more than 1/3 of the input stream. This is due to the fact that the interaction of the liquid flow through the axial hole wall with the main fluid tangential flow chamber is formed by twisting a centrifugal fluid jet stream. If shut off the flow baffle more than 1/ 3, then the interaction streams will form a liquid stream, which is typical for the ink jet nozzles so as to be commensurate with the flow commensurate with the angular momentum.

When placing the partition so that it covers less than 1/3 of the input stream, the amount of fluid motion from the main chamber tightening will significantly exceed the amount of motion of the fluid flowing out of the secondary camera twist. Although the total amount of momentum decrease because of interaction of flows, but it is characteristic of liquid outflow for swirler (hollow torch liquid spray).

The ratio of flows from the upper chamber of twist (optional) and lower (main) diameter hole defined in the partition wall 3, which must also be equal to 1/3 the diameter of the injector nozzle.

These conditions will allow the manufacture of the nozzle to form a liquid stream characteristic centrifugal spray nozzles with a high flow coefficient, ie low energy consumption for atomization.

## 3.3 Ejection device for sulphitation of sugar fluids

Sulphitation process is $SO_2$ removal processes of liquid (water or sugar solution) and subsequent chemical reaction with the substances contained in the sugar solution. The completeness of sulphitation (achieving the optimal pH of the solution) will depend completeness diffusing sucrose in water (sulphitation water) and quality of the resulting sugar (sulphitation juice and syrup).

Typical construction machines sulphitation sugar solutions have an apparatus, which consists of a separating vessel, and a cyclone connected thereto tangentially ejection device for interaction with liquid sulfur dioxide. The ejector is an elongated mixing chamber in the form of a pipe with a connection for the supply of sulfur dioxide and mounted on the end plate with openings for the liquid dispersion.

The disadvantage of this design of sulphitator is that, firstly, be used as working disk nozzle with apertures to disperse the fluid is ineffective. Secondly, after the working fluid and gas ejected by sulphitation fell into a mixing chamber, the flow passes stabilizing mixture. Updating minor phase contact surface, wherein the mass transfer rate significantly decreases (high mass flow rate condition is constant and intense contact of surface renewal phases) and to fully conduction, the process with a high coefficient of use $SO_2$ necessary to increase size of the device.

Perforated disc, which is used as a nebulizer (operating nozzle) does not create a large surface contact between the phases, because it is technically more jet nozzles.

Such atomizers are characterized by a compact jet (small opening angle of the spray cone of the liquid), so that its disintegration into droplets takes place at a great distance from the injector nozzle, which reduces the rate of mass transfer of gas into the liquid phase. In addition, low ejection coefficient, as the fluid interacts with the gas only the outside of the torch.

The first disadvantage is eliminated by using as a working fluid nozzle centrifugal, centrifugal jet nozzles or cavitation (we suggest using a centrifugal jet nozzle as the most simple and effective enough).

The second drawback is eliminated by creating conditions of turbulent flow liquid-gas emulsion in the mixing chamber, the collision of a liquid droplet, ie, with continuous updating phase contact surface.

Replacing the drive with the holes in the centrifugal, centrifugal or cavitation jet nozzles allowed qualitatively change the pattern of mass transfer. Through the nozzle of the injector fluid flows with large opening angle of the torch (40 ... 90 °) and breaks into droplets is almost at the nozzle of the injector. Atomized liquid is distributed evenly cross-section of the mixing chamber and creates a large surface contact between the phases, which is a condition of high mass transfer rates and high enough ejection coefficient.

Thus, the use as the working nozzle in the mixing chamber of sulphitator these types of nozzles leads to two positive effects:

- Significantly increases the surface contact between the phases by creating a large number of droplets of liquid that fills the entire cross section of the mixing chamber;

- Increases the ejection ratio of the gas phase.

However, after the initial contact of liquid droplets with sulphitation gas when the mass transfer rate is much higher than during steady-state, the movement is stabilized emulsion of a liquid - gas mixture along the chamber, the contact surface renewal phases slows and a decrease in the mass transfer rate of $SO_2$.

To be able to carry out the process sulphitation in intensive mode at constant updating phase contact surface sulphitator new design of [32].

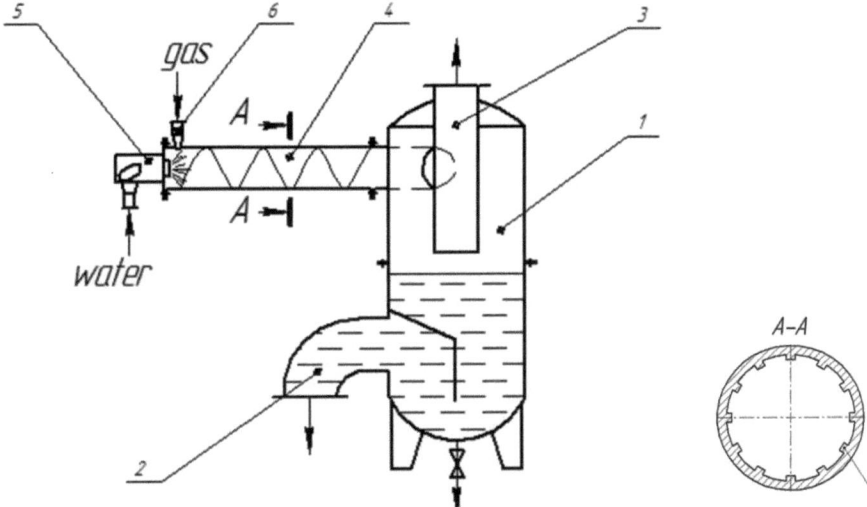

Fig. 8 Construction of sulphitator

Sulphitator (Fig. 8) separating tank consists of a cyclone 1 with a socket 2 of the treated liquid discharge pipe 3 and the removal of waste gas into the atmosphere. To the separating tank, 1 is connected tangentially to the mixing chamber as the pipe 4, and on the other side of the mixing chamber arranged coaxially centrifugal jet nozzle 5. Initially, a mixing chamber 4 is sulphitation feed pipe 6 in the gas apparatus, and in the mixing chamber 7, a guiding device formed as a helical cutting.

This design of the mixing chamber leads to the fact that part of the fluid that flows through the inner wall of a film and incident on a guiding device, it curls partially disrupted with sharp cutting edges of the screw. Next, change direction to the middle of the mixing chamber, where it reacts with the liquid droplets moving along its axis. Upon collision of liquid droplets passes intensive Update surface and thus accelerating the process of mass transfer. Thus, the helical cutting performed inside the mixing chamber is an additional baffle flow liquid - gas mixture. The surface of contact between the phases is continuously updated, the path of the liquid in the mixing chamber increases and, consequently, increases the contact time of finely dispersed liquid and gas increases the efficiency of mass transfer.

The device of the proposed design can be used as an effective sulphitator and mass transfer apparatus for carrying out mass transfer processes in the liquid - gas dust catcher, or as a mixer. In all cases, it will be increasing capacity of ejection of such devices (according to the literature and our preliminary studies increase ejection coefficient can be increased tenfold).

## 3.4 De ammonization of condensates

Rational use of water in a sugar factory provides for the return press water and ammonium condensates as an extraction for the diffusion process. Obstacle to the direct use of ammonium condensates is a high concentration $NH_3$, reaching 300 mg/l. This condensate must pass the proper treatment for reducing the ammonium content therein to 80 mg/l, whereby it can be used as part of the feed water for extraction. This reduces the need of the plant in fresh water barometric reduce discharges. In addition, the use of press water and ammonium condensates positive impact on the work itself and diffusion plant juice evaporation station. This is because such water has hardness, decontaminated by microorganisms. Diffusion juice, which is obtained by using ammonium condensates as part of the feed water for the extraction, has high purity to the juice, which is obtained by using as an extractant of barometric water. This is proved by experimental research and discussion on the use as part of the feedwater of ammonium condensate extraction of sugar from beet chips resolved in favor of its use.

Unresolved rational hardware design process of de ammonization with minimal energy cost.

Several flowsheets obtain de ammonization condensates [27], using as the de sorbent, both air and steam. The final concentration of ammonia, which is achieved in this case, is 50 - 80 mg/l. As the stripper can be used de ammonizators with nozzles for spraying ammonium condensates apparatus ebulating, desorption columns packed, electrodialysis plant for ammonium removal. There have been attempts to use as a stripper distillation column.

When carrying out the process condensates of de ammonization bubbling observed by the high cost of energy for heating the large volumes of air to an air heater to an optimum temperature and subsequent desorption of $NH_3$ bubbling through the liquid layer. At a rate of condensation on de ammonization [27] 5 m³/hr must submit air about 900 m³/hr, or one for every 180 m³ of condensate must m3 of air at a temperature of 80 $NH_3$ desorption process 80... $85^0 C$ .

During de ammonization ferry to reach a final concentration of ammonia in the water about 80 mg/l need about 3.2 - 3.3% of the third pair of evaporator housing.

Thus, with all the advantages of the use as feedwater of de ammonization condensates, the process by bubbling de ammonization enough energy intensive and requires large amounts of steam or compressed air.

The process of mass transfer from the aqueous solutions of $NH_3$ in the gas phase is described by the well-known equation [5]:

$$N_A = K_\Gamma F(y - y^*) \ , \qquad (20)$$

where $N_A$ - flow of desorbed component kmol / s;

$K_\Gamma$ - mass transfer coefficient, referred to the gas phase ( m / s);

$F$ - interface ;

$y$ - the concentration of ammonia in the gas ( kmol/m³ );

$y*$ - gas concentration equilibrium with the concentration in the liquid ( kmol/m³ ).

Mass transfer coefficient in turn is from the expression:

$$\frac{1}{K_\Gamma} = \frac{1}{\beta_\Gamma} + \frac{m}{\beta_\text{ж}}, \qquad (21)$$

where $\beta_\Gamma$ - mass transfer coefficient in the gas phase , m / s ;

$\beta_\text{ж}$ - mass transfer coefficient in the liquid phase , m / s ;

*m* - constant phase equilibrium.

In case of desorption are well soluble gases is assumed that the main resistance is in the gas phase and may be considered only a particular mass transfer coefficient in a gas phase.

However, due to the impossibility of taking into account all factors affecting the mass transfer (end effects , mass transfer during bubbling or spray mode in the system of polydisperse droplets , unequal in terms of mass transfer rate , etc.) to facilitate research use mass transfer coefficient referred to the unit volume of reactor :

$$k_v = K_\Gamma A, \tag{22}$$

where - $k_v$ volumetric mass transfer coefficient ;

$A$ - specific surface area of contact between the phases ;

$K_\textit{г}$ - coefficient of mass transfer in the gas phase.

Mass transfer equation in this case will be:

$$N_A = k_v V_{a\textit{б}} \, dc/d\tau \;, \tag{23}$$

where $V_{a\textit{б}}$ - volume absorber ( stripper );

$dc$ - the driving force of desorption ;

$d\tau$ - desorption time .

As for the driving force for the process of desorption of ammonia from the air in the liquid phase , then it is $NH_3$ partial pressure difference between the pressure of the ammonia and the equilibrium pressure in the liquid ammonia in the gas phase. Since ammonia is very soluble gas, the Henry's law is not subject, and to find the partial pressure of the aqueous solution (mm Hg. Tbsp.) Can use the formula:

$$\lg P_{NH_3} = -1750/T + 1.1 \lg M + 7 \;, \tag{24}$$

where $T$ - absolute temperature , $^\circ K$ ;

$M$ - $NH_3$ content in the solution kg-mol/m$^3$.

Ammonia amount transferred depends on the mass transfer coefficient, the surface of contact between the phases, the driving force for the process. Increase surface contact between the phases leads to an unambiguous increase in the desorption process. It is also known that the numerical value of the mass transfer coefficient depends on the hydrodynamic interaction of the phases (the numerical value of Reynolds number). Increasing the driving force for the process can be achieved by conducting the process in counterflow and increasing the speed of gas phase discharge, which is a part of desorbed ammonium.

In general, these findings are known from the analysis of the mass transfer equation. Feature is how to achieve the intensification of mass transfer. Sugar factories mainly adopted bubbling ammonia desorption method, which, as noted above, provides compression of large volumes of air and passing them layer ammonia condensates.

Promising way of intensification of mass transfer processes are:

- a significant increase in the surface contact by using a fine and monodisperse injector atomization fluid [28, 29, 30, 31];

- a significant increase in the mass transfer coefficient for creating hydrodynamic conditions of turbulence flow through the effective use of ejection apparatuses. When this is achieved by ejecting also sufficiently large volume of air or steam as the de sorbent without additional energy input.

These patterns and mass transfer mechanisms described desorption of ammonia allowed to create schemes in [33] speed of de ammonization condensates (Fig. 9).

As a first step used compact efficient ejection apparatus 1, which is a working nozzle, centrifugal - jet nozzle that allows you to fine and monodisperse droplets. In the mixing chamber, a high degree of dispersion flows updatable surface, which is a condition for the effective desorption NH3 in the air. For the subsequent separation of the air - water interface in the first stage cyclone is used de ammonization 2. Saturated with ammonium in ejection apparatus air through the chimney into the atmosphere or removed by vacuum capacitor.

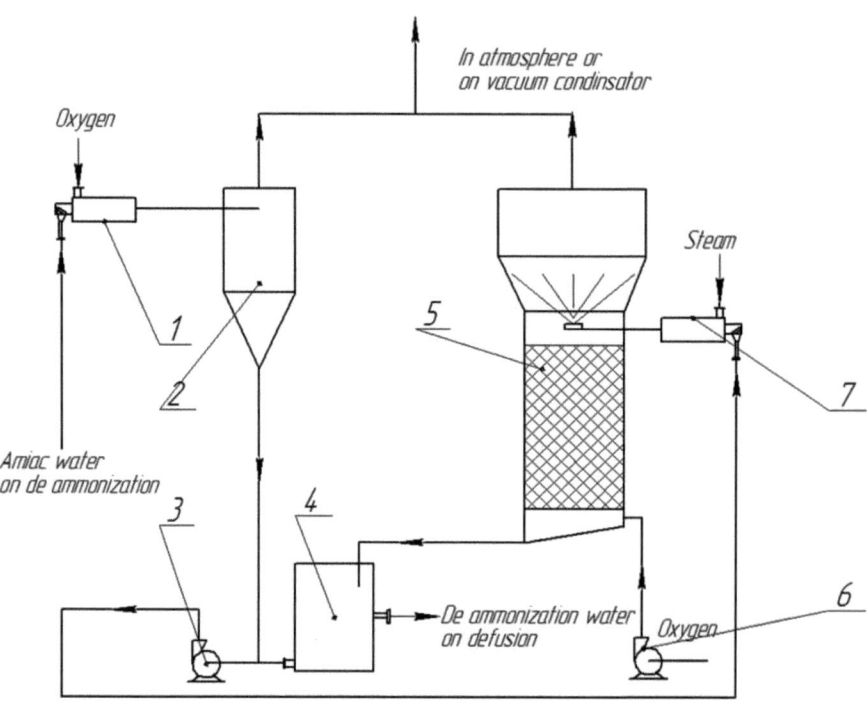

Fig. 9 Scheme of de ammonization condensates

Partially de ammonization water at a concentration of 180 ... 200 mg / l of the pump 3 is fed to the second stage of de ammonization, which is used as an ejection unit 7 and a packed column 5. The de sorbent is the fourth pair evaporator housing and the air that is pumped by low-pressure fan 6 to the bottom of the packed column under the distributor sieve.

Induction unit 7 is designed for desorption of ammonia vapor and maintain the optimum process temperature of de ammonization with low ejection potential fourth pair housing evaporator plant. This enhances the evaporators and condensers results in ammonia water to heating that is necessary for ammonia desorption in an optimum mode, as the temperature decreases $NH_3$ desorption conditions deteriorate. Phase separation takes place at the top of the packed column.

Packed column 5 consists of a cylindrical container, the bottom of which is located a sieve for air distributor, which is loaded on the attachment (for example, Raschig rings).

The fluid ejection device 7 after distributed at the top of the packed column and flows in the form of droplets, thin films, down and towards her fan 6 supplies air to which desorbs ammonium. In order to ensure the effect provided by the presence de ammonization fluid circulation circuit, which includes the collection of circulating 4, the bottom of which water is supplied to the circulating pump 3, which ensures a stable operation and the second-stage ejector de ammonization. From the middle part of the collection de ammonization water is supplied by diffusion.

Calculation stripper elements is carried out by known methods. So jets are calculated on the recommendations [28], the calculation of ejectors - according to [1]. The cap stripper calculated by [5] from the rate of suspension fluid into a column. The height of the packed bed is the specific load conditions phases.

This two-stage scheme of de ammonization condensates allows you to get a final concentration of ammonium in the water about 50 mg/l, which is acceptable for use in the diffusion process.

## 3.5 Saturation

Carbonation process in the sugar industry is for physic - chemical treatment of non-sugars from the sugar solution by treating the solution with a sugar content of 2.5 ... 3 % CaO carbonation gas with a concentration of about 35 % $CO_2$. As a result of the chemical reaction produces a calcium carbonate $CaCO_3$, which is adsorbed on the surface of nonsugar [26, 27].

Processing sugar solution carbonation gas is usually carried out in the apparatus at the height of bubble-type liquid layer of about 4 meters above process in a sugar factory is one of the most energy-intensive and at the same time, the main sugar in the cleaning solution from the non-sugars. The process is carried out in two stages.

The main disadvantages of the first and second sets of saturation are:

- low efficiency of purification of non-sugars from the sugar solution, which does not exceed 40 % ;

- low efficiency of utilization of carbon dioxide ( $CO_2$ utilization factor in the first carbonation apparatus - 60-65 % , in a second carbonation devices - not more than 50 %) ;

- emission of exhaust gas carbonation lead to significant air pollution;

- a significant loss of heat from the exhaust gas.

One reason for the low efficiency of purification of sugar solution from non-sugars in the apparatus is a saturation phenomenon of "total alkalinity", ie Mix juice servings almost treated to a final value of alkalinity and fresh portions of juice that arrive at the machine on saturation. For high-quality processing of sugar solution should be a zone of primary nucleation of calcium carbonate crystals and the subsequent growth and maturation zone with simultaneous adsorption of non-sugars on the surface of a continuously updated.

Increase cleaning efficiency carbonation lime juice 1 - carbon-dioxide is achieved by partitioning the machine. In the transition section from a section gradually decreasing alkalinity of juice and nonsugars precipitated calcium carbonate crystals at optimum alkalinity.

To increase the use of carbon dioxide in the carbonation devices available in two ways:

- raising the level of juice, which is bubbled through the carbonated gas or installation in a bubble volume carbonator gas distributors of various designs, pulsators;

- installation in space between carbonator additional distributors juice;

- well - known and practically used method of increasing the use of $CO_2$ gas as carbonated before being fed into the apparatus saturated steam to an equilibrium state or partial exceeds its saturation temperature [25].

Increasing the use of carbon dioxide carbonation gas by increasing the level of juice, which is bubbled through the carbonated gas [27] does increase the utilization rate of $CO_2$. For example, increasing the level of juice from 4 m to 6 m, the utilization factor increases from 65% to 90 %.

One of the effective ways to increase the use of $CO_2$ is the use of hydraulic nozzles to disperse the juice in space of carbonator. When this is achieved by utilization of $CO_2$ by 15 - 20 % higher [27].

The advantage of this method of increasing the use of $CO_2$ is minor capital costs and the ability to force the plant to make upgrades to existing carbonator.

We estimate the energy costs of these two ways to increase the use of $CO_2$.

Increased saturation juice machines have a fairly simple solution to achieve a higher utilization of $CO_2$. However, addition of new capital installation costs significantly more powerful hardware increases energy costs for gas compression. Determine the work of adiabatic compression of 1m3 gas Lad expended for additional saturation by raising the level of juice in the machine with 4 m (min overpressure gas compression in the compressor of 0.04 MPa) to 6 m (min overpressure compression - 0.06 MPa) when the inlet gas temperature of 30°C [34]:

$$L_{a\partial} = \frac{k}{k-1} \cdot p_1 \cdot v_1 \left[ \left( \frac{p_2}{p_1} \right)^{\frac{k-1}{k}} - 1 \right],$$ (25)

where $k$ - the adiabatic index;

$p_1$ - the suction pressure, $p_1 = 0.03$ MPa;

$p_2$ - discharge pressure, with the possibility of injecting gas bubbling through the liquid layer 4 m $_{p_2}$ $_1 = 0.14$ MPa, the discharge gas to allow bubbling through the liquid layer 6 m $_{p_2}$ $_2 = 0.16$ MPa;

$v_1$ - specific volume of gas at suction; $v_1 = 0.714$ m³/kg.

An estimate shows that the compressive each 1m³ of gas under these conditions need to spend an additional 0.01 MJ of energy.

Motor shaft Compressive compressor 1 m³ of gas and at its mechanical efficiency can be determined by the formula:

$$N = \frac{L_{a\partial}}{\eta_{\text{мех}}} = \frac{p_1 v_1 \ln\left( \frac{p_2}{p_1} \right)}{1000 \eta_{\text{мех}}}$$ (26)

After substituting the value of the power N, when the gas is compressed to a pressure of 0.04 MPa excess amount:

$N_1 = 42.6$ КВт

and when the gas is compressed to 0.06 MPa overpressure:

$N_2 = 52.8$ КВт

That is, to compress 1 m³ of gas at the same initial conditions to a final pressure of 0,06 MPa, which is the minimum necessary for the carbonation bubbling gas through the juice 6 m compared with bubbling method saturation at 4 m must also spend 10,2 kW.

Gas temperature adiabatic compression is given by:

$$T_{\kappa} = (273 + t_1)\left(\frac{p_2}{p_1}\right)^{\frac{k-1}{k}}, \qquad (27)$$

where $t_1$ - inlet gas temperature.

In the first case, the gas will have a temperature of about $83^0C$ in the second - $96^0C$.

When using spraying juice space carbonator pump his power is determined from the expression:

$$N = \frac{Q\rho g H}{1000\xi}, \qquad (28)$$

where - $Q$ - flow rate , $m^3$ / s;

$H$ - pressurized fluid , m;

$\xi$ - efficiency a pump power of 0.8.

To be able to disperse the fluid nozzle its pressure in the supply line should be about 0.2 MPa.

Substituting numerical values gives the necessary indicative power supply under 1 $м^3$ of liquid - 245 kW.

Material balance of carbon dioxide in view of its utilization for the treatment of each cubic meter of sugar solution at the first saturation takes about forty cubic meters carbonation gas. Then, to get the utilization of carbon dioxide carbonation gas about 80 % by increasing the level of juice in the machine up to 6 m or by atomization of sugar solution in the saturator space, in the latter case, the cost of energy will be about 7 times less.

The cleaning effect of the sugar solution at atomization juice in carbonator space increased by partitioning the machine (the upper part of the saturator works in spray mode, the bottom - in a bubble). This leads to the separation zone of higher alkalinity (spray saturator part), where the formation of crystalline calcium carbonate of high adsorption capacity, thus achieving a higher quality of purification.

By increasing the level of juice in the saturator to achieve higher utilization of $CO_2$ only increases the amount of vehicles, hydrodynamic conditions deteriorate , there is the phenomenon of " total alkalinity " (alkalinity juice machine juice alkalinity is at its output) and the expected significant improvement of quality of the juice is not observed.

Despite the undeniable advantages, atomization juice between space in front of a simple increase in the level of juice in accordance with the assumptions made in choosing ways to improve the utilization of $CO_2$ by plants is taken the second path. Obviously, here plays the role of the usual hundred years of experience in a bubble station of saturation mode, predefined shortcomings of equipment and ways to combat them.

In Russia's recent work is underway on the use of ejection apparatus as the first stage of saturation. Such devices are simple in design, effective for the mass transfer processes, reduce the pressure of carbonation gas supply to the unit partition the Saturators and permit sugar solutions containing less non-sugars in the formation of calcium carbonate with high adsorption capacity.

Saturator such design [3] is a two-piece unit with the first stage and the subsequent ejection of the bubble. In operating the nozzle ejection apparatus served defecated juice and a portion (about 40 % - 60 %) fresh carbonation gas. The second part of the gas enters the bubble carbonator tank model. The authors also provide for other schemes of saturation in the first carbonation devices using ejectors.

Thus achieved increased utilization of $CO_2$, such as sugar saturation - raw gas concentration at the initial 28 - 36 % to 63-72 %. Carbonation time was 1 min. and an ejection step in 4 min. in a bubble.

Known technical solutions to increase the use of carbon dioxide mainly concern a single unit. Integrated solutions use the exhaust gas of the carbonation devices first and second no carbonation.

Of interest is another method of increasing the use of carbon dioxide, based on the reuse of waste carbonation gas. However, implementation of such a method inhibits the fact that compression of the hot and moist gas requires considerable energy commensurate with the positive effect from such reuse.

Simple material balance calculations for carbon dioxide indicates that the exhaust gas from the carbonation of the first carbonation unit $CO_2$ concentration is high enough and sufficient for use of this gas in the second carbonatation juice at a ratio - carbonated gas is 1/20.

Since carbonated gas after the first carbonation, juice is at a temperature and saturated with water vapor, the use of this gas in the second carbonation will reduce heat loss and reduce exhaust gas temperature drop of juice on the second saturation, thus eliminating it from further heating. In addition, the reuse of carbonation gas saturation after the first apparatus to the second carbonation lead to a fuller use of the carbon dioxide that will reduce air pollution. An important fact is the fact that decrease the overall consumption of carbonated gas to process the first and second carbonation.

Solve the problem of re-use carbonation gas ejection can use the machine as a pump.

We propose a new method of carbonation sugar solutions, which is based on the use of the exhaust gas of the first carbonated carbonation as the second carbonation gas saturation [35]. A method of purifying the sugar solution is the following (a variant flowsheet) - Fig. 10.

Preliming sugar solution gets into the main liming 1 defecation, which also added to 2.5 - 3% CaO where the chemical treatment of non-sugars of the sugar solution. Alkalinity of the juice with a pH of 12.2 - 12.3.

Limed sugar solution fed into the unit first 2 saturation, where he handled the carbonation gas with a concentration of 30 - 35 % $CO_2$, which is produced at a plant in lime- kilns. In the saturator at about $78^0C$ of the process as a result of carbon dioxide absorption, and chemical reaction with the dissolved lime calcium carbonate crystals are formed which are adsorbed on their surfaces and thus nonsugar occurs physico - chemical treatment.

After treatment with alkaline solution to a final 0.1 % CaO juice is heated in preheater 3 to a temperature of 85 … $90^0C$ and to the discharge collector 4 , which is fed by gravity from the filters in 5 thickeners (or sedimentation ), where the separation of the clarified juice fraction and slurry. The thickened slurry sent to the vacuum filters or other structure (chamber filter presses) to select six of her clarified juice.

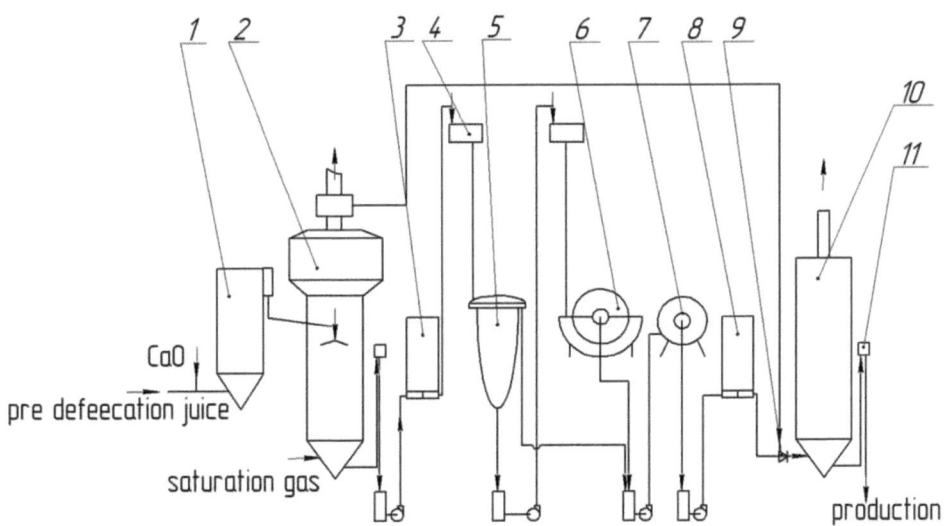

Fig. 10 Flowsheet use $CO_2$

The whole of the clarified juice fed to the control filtering disc filters 7.

After the juice is pumped to the filter by the heater 8 is heated to a temperature of $95^0C$ and routed to ejection mixing apparatus 9. This is also supplies, when necessary, to increase the adsorption treatment and the precipitate filtered obtaining milk of lime in an amount of about 0.2 % CaO. In ejection apparatus, ejection occurs carbonation gas exhaust from the machine first carbonation.

Coming up with the ejection apparatus juice- emulsion gas enters the apparatus 10 to the second carbonation, where alkalinity juice reduced to a final 0,015 - 0.02% CaO and have the juice through the overflow box 11 enters for further processing in the production.

In the carbonation gas, which enters the ejection apparatus of the first device saturation, there is a sufficient amount of carbon dioxide for carbonation process of the second. Since the initial concentration of $CO_2$ in the first saturation unit 30 - 35 % and the utilization factor of carbon dioxide of about 60 - 70 % final concentration of $CO_2$ in the gas is 9 - 14%.

At such concentrations of reagents for the normal process of the second carbonation juice ratio should be performed gas about 20 (ie, each part juice should eject spent 20 pieces of apparatus carbonation gas first carbonation).

Such ejection coefficient obtains in the proposed jet ejection apparatus with dispersed jet, where the working nozzle sprays used various liquids (nozzles).

Carbonated gas that comes from the limekiln in the first carbonation machine saturates with water vapor at a temperature of about 30 $^0$ C and has a temperature of about 70 $^0$ C after compression. In saturator, gas saturated with water vapor and comes to equilibrium at 75 – 78 $^0$ C. The temperature of the juice reduces by 2 … 5 $^0$ C. In operation, the sugar solution purification scheme adopted for the type, the concentration of carbonated gas $CO_2$ 9 … 14% is removed in the atmosphere, polluting it, and greatly with taking a considerable amount of heat.

In the case of purification of the sugar solution proposed scheme in a second carbonation unit, where the process temperature is $95^0$ … $102^0$ C to bring the carbonation gas in equilibrium with a liquid heat require considerably less than when using a gas with a lime kiln, and hence reduction temperature of juice in the second unit will be minimal carbonation.

Thus, for operation of the plant according to this scheme is additionally installed with the ejection unit of the dispersed stream of liquid as the working fluid supply and exhaust conduits carbonation gas of the first carbonation unit to the ejection apparatus.

Use of waste gas from the carbonation unit first carbonation will reduce the total consumption of carbonated gas to hold carbonation processes , and thus will save fuel for carbonation gas in the lime- kilns .

Also of interest is the use of the flue gas from the second unit to the first carbonation. The scheme [36] proposed to introduce a sugar factory (Fig. 11).

Processing sugar solutions is mainly liming 1, the first stage of the first saturator 2, the second stage of the first carbonator 3, in the second carbonation unit 4. Defecated juice on the first stage of the first carbonation is supplied by a pump 5.

In the second stage of the first unit and the second processing solution saturation passes carbonation gas, which is produced - lime kilns.

According to this scheme assumes partitioning apparatus first carbonation, as in the previous case, with ejection apparatus. The simultaneous formation of nucleation of calcium carbonate in the ejection stage of the saturator and the subsequent growth of the bubble allows the juices to obtain a higher picture quality than the processing of a typical bubbling juice saturator, where the phases of nucleation and subsequent growth occur simultaneously.

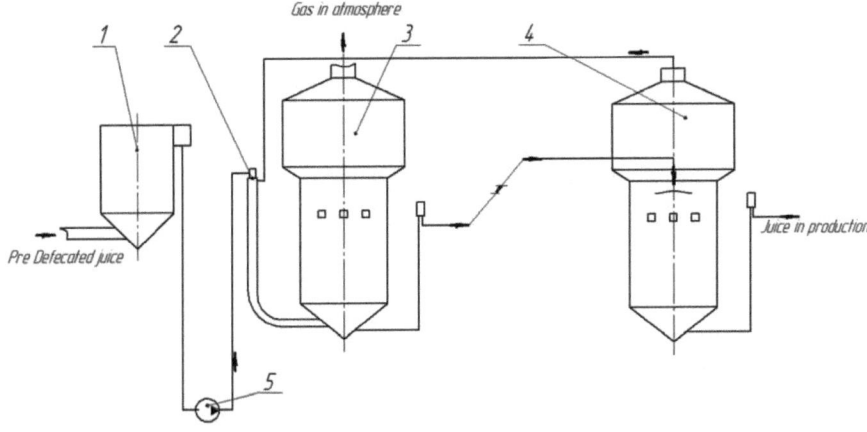

Fig. 11 Flowsheet use waste carbonation gas saturation of the second unit

The second saturation occurs at high temperatures (100 ... $102^0$ C) and a low coefficient of utilization of carbon dioxide (50...60 %), so the exhaust carbonated gas is saturated with water vapor at a given temperature and has a high concentration of $CO_2$.

When using such a gas to the first stage of the first carbonation that extends at a lower temperature (75 ... $80^0$ C), heat transfer will occur and sugar solution will be no need of heating the juice prior to filtration. The presence of carbon dioxide , about 14% , enough for the first stage of the first carbonation 20...30 - percentage degree of carbonation solution , ie lime which is in a dissolved state , it is necessary for the formation of calcium carbonate crystals of high adsorption capacity .

Wherein carbonation gas is used which is substantially less than when using a saturator at a typical circuit, and hence pollution of the atmosphere with carbon dioxide will be lower. Use of waste gas from the carbonation unit second carbonation will reduce overall costs to conduct carbonation gas saturation processes.

Most economical machine for the first stage of saturation is the ejection device having a dispersed stream of working fluid. When using such a device is achieved as follows:

- Ejection coefficient of such a device is quite high, which will use the entire volume of the exhaust gas of the carbonation unit second carbonation and prevent the release of hot gas into the atmosphere;

- Do not require additional energy consumption for compression spent carbonation gas for its usage in the first stage of the first carbonation;

- The developed surface of contact between the phases in the ejection apparatus with dispersed jet is one of the conditions of rapid saturation, guarantee a high degree of cleaning solution from the non-sugars.

# Literature

1. Лямаев, Б.Ф. Гидроструйные насосы и установки. / Б.Ф. Лямаев. - Л.: Машиностроение. Ленинград. отдел., 1988. - 256 с.

2. Гребенюк, С.М. Технологическое оборудование сахарных заводов. / С.М. Гребенюк. – М.: КолосС, 2007. – 520 с.

3. Воинов, С.К. Совершенствование способа инжекционно-барботажной сатурации клеровки сахара-сырца: дис. ... канд. техн. наук: 05.18.05 / Воинов Сергей Константинович. – МГУПП. – М., 2008. – 162 с.

4. Шервуд,Т. Массопередача / Шервуд Т., Пигфорд Р., Уилки Ч. Пер. с англ. – М.: Химия, 1982. – 696 с.

5. Рамм, В.М. Абсорбция газов / В.М. Рамм изд. 2-е, переработ. и доп. М.: Химия, 1976. - 656.с.

6. Броунштейн, Б.И. Гидродинамика, массо,- и теплообмен в дисперсных системах / Б.И. Броунштейн, Г.А. Фишбейн - Л.: Химия, 1977.- 280 с.

7. Кафаров, В.В. Основы массопередачи / В.В. Кафаров - 2-е изд., перераб. и дополн. - М.: Высшая школа, 1972.- 496 с.

8. Савельев, Н.И. Метод расчета эффективности массопереноса в прямоточно-вихревых контактных устройствах ректификационных и абсорбционных аппаратов / Н.И. Савельев, Н.А. Николаев, В.А. Малюсов - ТОХТ, 1981, т. ХУ, № 5, с. 643 - 649.

9. Броунштейн, Б.И. Гидродинамика, массо- и теплообмен в колонных аппаратах / Б.И. Броунштейн, В.В. Щеголев – Л.: Химия, 1988. – 336 с. – ISBN 5-7245-0100-7.

10. Kroepelek H., Neuman H.M., Prot E. // Erdol u. Cohle 1959. Bd.12 № 5. S. 344 – 347.

11. Ilcowich D.//Colin. Czech.Chem.Communs. 1934.В. 6. s. 498.

12. Бутвин, А.Н. Исследование массопередачи в системе падающая капля - фтористый водород / А.Н. Бутвин, А.А. Эннан, В.М. Солодов, Г.Г. Михайленко // Тез. докл. 2-е Всесоюзное совещание по проблеме "Абсорбция газов", ГИАП.- Черкассы, 1983, часть I.- с. 129-132.

13. Плитт, И.Г. К теории хемосорбции в прямоточных потоках газа и капель "большого" диаметра / И.Г. Плитт. - ЖПХ, 1965, № 7,-с. 1527-1536.

14. Кабаков, М.И. Исследование гидродинамических неоднородностей в скруббере большого диаметра / М.И. Кабаков, А.С. Стелько, А.М. Розен. - ТОХТ, I985, т. XIX, № 3, с. 403- 405.

15. Кисилев, Е.И. Расчет гидравлической обстановки в распылительном аппарате с вторичным дроблением жидкости / Е.И. Кисиле, Д.О. Бытев, А.М. Зайцев, А.А. Дозоров // Тез. докл. Всесоюзная научная конференция " Повышение эффективности, совершенствование процессов и аппаратов химических производств", ХПИ. - Харьков, I985, часть 3. - с. 59 - 60.

16. Савельев, Н.И. Влияние диаметра капель на массоперенос в прямоточно-вихревых контактных устройствах / Н.И. Савельев, Ю.Н. Бодров, Н.А. Николаев // Тез. докл. 2-е Всесоюзное совещание по проблеме "Абсорбция газов", ГИАП.- Черкассы, I983, часть I. - с. 133 - 135.

17. Холин, Б.Г. 0 дроблении капель в вихревых противоточных массообменных апаратах / Б.Г. Холин, В.А. Кравченко, В.И. Склабинский // Тез. докл. Всесоюзная научная конференция "Повышение эффективности, совершенствование процессов и аппаратов химических производств", ХПИ. - Харьков, 1985. - с. 52 - 53.

18. Выскребцов, В.Б. Исследование абсорбции при полидисперсном распыливании жидкости и разработка эфективных сульфитационных аппартов сахарного производства: - дис. ... канд. тех. наук: 05.18.12 / Выскребцов Владимир Борисович – КТИПП. - Киев, 1978. - 159 с.

19. Выскребцов, В.Б. К математической модели абсорбции в распылительных колоннах / В.Б. Выскребцов, Л.В. Леонтович, И.М. Федоткин, М.И. Павлищев - ТОХТ, 1977,т. XI, № 6, - с. 812 - 824.

20. Кафаров, В.В. Основы массопередачи / В.В. Кафаров - 2-е изд., перераб. и дополн.- М.: Высшая школа, I972.- 496 с.

21. Василенко, С.М. Основи тепломасообміну : підручник / С.М. Василенко, А.І. Українець, В.В. Олішевський. / за ред.. акад.. УААН І.С. Гулого - К.: НУХТ, 2004. – 250 с. – ISBN 966-612-030-5.

22. Рева, Л.П. Напрямки комплексного удосконалення сучасних технологічних процесів очищення дифузійного соку. / Л.П. Рева, В.А. Шостаковський, Т.І. Антоненко.// Цукор України, 2007, № 5 – 6. - с. 12 – 17.

23. Выскребцов, В.Б. Производственные испытания распылительного сатуратора под давлением. / В.Б. Выскребцов, В.В. Пономаренко, В.И. Бочкин. // Сахарная промышленность, 1986, №9. - с. 30 – 32.

24. Рид, Р. Свойства газов и жидкостей / Р. Рид, Дж. Праусниц, Т. Шервуд / Пер. с англ. Под ред. Б.И.Соколова. - 3-е изд., перераб. и дополн. – Л.: Химия,1982. – 592 с. – Нью-Йорк, 1977.

25. А.с. № 1723133 (СССР), МКИ 3 С 13 D 3/04. Способ очистки сахаросодержащих растворов / В.Б.Выскребцов, В.В.Пономаренко, Л.В.Пихоцкий (СССР).-SU №1723133 A 1; заявл.11.07.89; опубл.30.03.92, Бюл. № 12.- 6 с.: ил.

26. Сапронов, А.Р. Технология сахарного производства / А.Р. Сапронов. - М.: Агропромиздат, 1986. - 431с.

27. Современные технологии и оборудование свеклосахарного производства. Ч.1 / В.О.Штангеев, В.Т.Корбер, Л.Г.Белостоцкий и др.; под ред. В.О.Штангеева - К: Цукор України, 2003 – 352 с. – ISBN 966-96351-0-1.

28. Выскребцов, В.Б. Об особенностях выбора рациональных размеров регулируемой центробежно-струйной форсунки для распыливания жидкостей и суспензий. / В.Б. Выскребцов, В.В. Пономаренко.– Рук. деп. ред. журн. «Известия вузов. Пищевая технология». – Деп. в ЦНИИТЭИлегпщемаш 2.12.1987, № 817 – мл.

29. А.с.№1382499. СССР, МКИ 3 В 05 В 1 / 34. Форсунка для распыливания жидкости. / В. Г. Выскребцов, В.Б. Выскребцов, В.В. Пономаренко (СССР). – 4152051/31-05; опубл. 23.03.88, Бюл. № 11. – с. 4.

30. Патент 99671 UA, МПК B05B 1/34 (2006.010). Форсунка / Пономаренко В.В. ; заявник Національний університет харчових технологій - № a201103095; заявл. 6.03.2011 ; опубл. 10.09.2012, Бюл. №17/2012.

31. Патент 45125 UA, МПК B05B 1/34 (2006.01). Форсунка для розпилювання рідини. / Пономаренко В.В., Петренко В.О.; заявник Національний університет харчових технологій - № u200905410; заявл. 29.05.2009 ; опубл. 26.10.2009, Бюл. №20/20019.

32. Патент 102782 UA, МПК C13B 20/10 (2011/01), B01F 3/04 (2006.01), B04C 5/04 (2006.01) Сульфітатор / Луговська О.А., Пономаренко В.В., Хитрий Я.С.; заявник Національний університет харчових технологій - № a2012 06202; заявл. 23.05.2012; опубл. 12.08.2013, Бюл. №15/2013.

33.    Патент 102419 Україна, МПК С 13 В 25/00 (2013.01) Спосіб деамонізації конденсатів цукрового виробництва / Пономаренко В. В.; Вискребцов В.Б.; власник: Національний університет харчових технологій. — а № 2011 08484; заявл. 06.07.11; опубл. 10.07.13, Бюл. № 13 – 5 с.

34.    Шлипченко, З.С. Насосы, компрессоры, вентиляторы. / З.С. Шлипченко, К., Техніка, 1976. - 368 с.

35.    Патент № 99474 UA, МПК С13В 20/06 (2011.01) Спосіб очищення цукрових розчинів / Мирончук В.Г., Пономаренко В.В., Гандабура І.В.- власник: Національний університет харчових технологій. — а № а201003097; заявл. 18.03.2010; опубл. 27.08.2012, Бюл. № 16 – 5 с.

36.    Патент № 82478 UA, МПК С13В 20/00 (2013.01). Спосіб сатурації цукрових розчинів. / Пономаренко В.В., Пушанко Н.Н.; власник: Національний університет харчових технологій. — u 201213587. заявл. 27.11.2012; опубл. 12.08.2013, Бюл. № 15 – 5 с.

Printed by Books on Demand GmbH, Norderstedt / Germany